The Universal Core Information Exchange Framework

Assessing Its Implications for Acquisition Programs

Daniel Gonzales, Chad J. R. Ohlandt, Eric Landree,
Carolyn Wong, Rima Bitar, John Hollywood

Prepared for the United States Navy

RAND NATIONAL DEFENSE RESEARCH INSTITUTE

The research described in this report was prepared for the United States Navy. The research was conducted within the RAND National Defense Research Institute, a federally funded research and development center sponsored by the Office of the Secretary of Defense, the Joint Staff, the Unified Combatant Commands, the Navy, the Marine Corps, the defense agencies, and the defense Intelligence Community under Contract W74V8H-06-C-0002.

Library of Congress Control Number: 2011926383

ISBN: 978-0-8330-5092-2

Published 2011 by the RAND Corporation
1776 Main Street, P.O. Box 2138, Santa Monica, CA 90407-2138
1200 South Hayes Street, Arlington, VA 22202-5050
4570 Fifth Avenue, Suite 600, Pittsburgh, PA 15213-2665
RAND URL: http://www.rand.org/
To order RAND documents or to obtain additional information, contact
Distribution Services: Telephone: (310) 451-7002;
Fax: (310) 451-6915; Email: order@rand.org

Preface

This report presents observations from an ongoing research project that is tasked with assessing and improving Department of Defense (DoD) and Navy policy for command, control, communications, and intelligence and for weapon programs. This report examines a new information exchange standard, Universal Core (UCore), its relationship to DoD data strategy and policy, its implementation options, and related technical issues that should be resolved prior to the widespread adoption of this powerful new interoperability mechanism. This research should be of interest to members of the Navy and of the broader DoD responsible for formulating, reviewing, or implementing DoD interoperability policy. It should be of particular interest to Navy program managers responsible for the development of information technology and national security system programs.

This research was sponsored by the Assistant Secretary of the Navy, Research, Development and Acquisition Chief Systems Engineer (ASN RDA CHSENG), and by the Office of the Secretary of Defense. It was conducted within the Acquisition and Technology Policy Center of the RAND National Defense Research Institute, a federally funded research and development center sponsored by the Office of the Secretary of Defense, the Joint Staff, the Unified Combatant Commands, the Navy, the Marine Corps, the defense agencies, and the defense Intelligence Community.

For more information on the RAND Acquisition and Technology Policy Center, see http://www.rand.org/nsrd/about/atp.html or contact the director (contact information is provided on the web page).

Contents

Figures

Tables

Summary

UCore 2.0 is an Extensible Markup Language (XML) schema designed for transmitting situational awareness data, which is applicable to a very broad range of data types. DoD, the Department of Justice (DoJ), the Department of Homeland Security (DHS), and the Office of the Director of National Intelligence (ODNI) jointly developed UCore to improve interoperability within and between U.S. government agencies.

UCore Has Promise

On the surface, UCore is simply another XML-based data standard available to the DoD acquisition community for establishing interoperability between DoD information systems. However, the data-standard "wrapping" and extensibility capabilities inherent in UCore give it the potential to significantly improve interoperability between DoD information systems and provide a new way to realize the promise of the DoD data strategy. In addition, UCore has the backing and development resources made available by a high-level design consensus between DoD, DHS, DoJ, and the Intelligence Community (IC).

An important aspect of the DoD data strategy is understandability, which requires both semantic and syntactic elements to achieve a working data model. Widespread adoption of a small set of common syntactic or XML standards, such as UCore, will improve understandability, especially for the unanticipated user. However, UCore is not a complete data model and by itself will not meet the interoperability needs of the Navy or the broader DoD. Extensions to UCore, such as Command and Control Core, are needed to make the schema more useful for Navy and other DoD users. The extensibility of UCore makes this increased usefulness possible.

Recognizing the potential benefits of UCore for DoD interoperability, the DoD Chief Information Officer (CIO) is advocating for accelerated adoption of UCore by acquisition programs through policy directives, instructions, and memoranda.

UCore Policy Is Immature

Most existing DoD policy relevant to the DoD data strategy does not mention UCore, and new DoD policy (especially Chairman of the Joint Chiefs of Staff Instruction 6212.01E) provides ambiguous and confusing guidance on UCore implementation requirements. Establishing a clear-cut DoD policy on UCore is complicated by the fact that UCore is neither a simple XML standard nor a complete data model that can meet all of DoD's interoperability requirements. Since UCore fits neither paradigm completely, stakeholders in UCore policy are likely to interpret UCore differently—as a standard or a data model or something in between. This creates policy challenges, since the authority to generate policy and guidance on UCore depends on how UCore is defined.[1]

UCore Pilot Projects

Alpha and beta testing of UCore 2.0 is complete, and UCore 2.0 was released for use in March 2009. Although UCore 2.0 demonstrates technical innovation and great potential for improving DoD interoperability, especially for the unanticipated user, there is little programmatic guidance for implementing UCore, and there are effectively no hard data on UCore bandwidth demands or cost implications. Bandwidth concerns are potentially significant because metadata tagging can increase message size and associated bandwidth requirements by more than an order of magnitude. Although such an increase in bandwidth demand may not be an issue in the high-capacity networks of the Global Information Grid, it could be problematic for low-bandwidth tactical-edge wireless networks.

At most, five of the 19 documented pilot projects assessed UCore performance in tests that actually used a DoD network. Two pilot projects appear to have been tested on a local area network, another two appear to have been tested on the Defense Information System Network, and one was tested on an unspecified network. None appear to have been tested on a tactical network.

These UCore pilot projects are not well documented, and there is not sufficient technical information to help acquisition program managers make informed UCore implementation decisions. Furthermore, detailed technical data on UCore implementation options, including information on UCore message sizes, are not readily available, even for only the minimal implementation of UCore. Therefore, this body of evidence cannot be used to assess the impact of UCore on Navy networks or to assert that UCore can be implemented effectively and without risk on Navy networks with bandwidth limitations.

UCore Implementation Options and Risks

The are several ways to implement UCore. Some of these implementation options are feasible only if extensions to UCore are available. Others can be implemented more quickly if existing XML schema are wrapped in UCore. Limited guidance and supporting technical information that describe the potential advantages and disadvantages of these implementation options, their costs, and their impact on the Navy networks are available.

The limited data on UCore message sizes and cost implications should be a source of concern in the DoD acquisition community, especially in light of UCore's flexibility and the variety of implementation methods available, both of which greatly complicate the technical and cost assessment of UCore systems.

XML messaging is typically verbose, and UCore is no exception. Additionally, the extensibility and wrapping inherent in UCore messages make the assessment of network impact

---

[1] For an in-depth exploration of information technology roles and responsibilities in the DoD, see Gonzales et al., 2010b.

even more difficult because there is significant potential variability in UCore message sizes. Finally, the cost of implementing UCore will vary greatly depending on specific implementation choices. These technical issues need to be understood and addressed before UCore implementation can be mandated.

Recommendations

In light of these findings, we make the following recommendations:

- The Navy should not require widespread UCore implementation at this time.
- DoD should undertake additional UCore piloting efforts to quantify potential system and program impacts. It should
  - Capture UCore vocabulary size, message sizes, system processing speeds, and network bandwidth consumption.
  - Model and document program UCore implementation costs for specific UCore implementation options.
- Any future UCore requirements or mandates should not originate ab initio from policy but should rather be informed first by the evaluation of technical and cost data from well-documented future pilot projects.
- Guidance for setting UCore requirements in program Initial Capabilities Documents should include an analysis of alternatives that weighs interoperability benefits against bandwidth impact. For example, systems that will regularly interact with DoJ DHS/IC systems clearly need to implement UCore or an equivalent, but tactical-edge systems with limited bandwidth might be forbidden to use UCore, and headquarters intelligence databases attached to wideband networks that accumulate data from the tactical edge might be allowed to generate UCore messages for passing information up the chain of command on high-bandwidth networks.
- The Navy should permit UCore implementation if (1) the program depends only on high-bandwidth networks or (2) UCore will not degrade or otherwise affect required real-time performance requirements for the system or system of systems to which the subject system belongs.
- The Navy should develop guidance for preferred UCore implementation approaches that accounts for different Navy networks and operational environments.
- DoD should develop a new, streamlined version of Ucore (UCore 3.0) that contains a small and extensible core vocabulary for the UCore core primitive data frameworks for "who" and "what."

Future Work

Related RAND Corporation research for the Director of Joint Interoperability, Under Secretary of Defense for Acquisition, Technology and Logistics, has shown that XML compression can greatly reduce the size of XML-tagged messages. This may enable UCore messaging to be used on networks that have bandwidth limitations. Further research is required both to determine how effective XML compression is when applied to UCore messaging and to determine which Navy network types could support compressed UCore messaging.

One of the greatest long-term potential benefits of UCore adoption will be the ability of different communities of interest to independently develop improved data tagging using UCore or a UCore extension *while remaining interoperable* (to a certain degree). However, to realize that capability, the UCore data model needs to avoid becoming a centrally managed affair—a situation that could inhibit innovation (as has occurred in the case of the National Information Exchange Model). DoD and the Navy should develop alternative ways of managing the UCore data model that encourage program experimentation while ensuring that bandwidth issues and implementation costs are addressed.

Acknowledgments

This report would not have been possible without support from the ASN RDA CHSENG office. The authors wish to specifically thank Cheryl Walton, Director of ASN RDA CHSENG's Standards, Policy and Guidance Directorate, for her guidance. We greatly appreciate the review of an early version of the manuscript and feedback provided by Jacqueline Knudson of the Defense Information Systems Agency. We also thank federal co-leads Dan Green and Olithia Strom of the UCore Development Team for sharing information on their program and insights on DoD data strategy, UCore, and related issues.

Finally, we thank Bob Anderson and Tony Hearn of RAND, the peer reviewers of this report, and Phil Antón, the director of the RAND Acquisition and Technology Policy Center, for their careful review of the manuscript.

Abbreviations

ASN	Assistant Secretary of the Navy
C2Core	Command and Control Core
C4I	command, control, communications, computers, and intelligence
CHSENG	Chief Systems Engineer
CIO	Chief Information Officer
CJCSI	Chairman of the Joint Chiefs of Staff Instruction
CoI	community of interest
CPD	Capability Production Document
CSG	carrier strike group
DHS	Department of Homeland Security
DISN	Defense Information System Network
DoD	Department of Defense
DoDD	Department of Defense Directive
DoDIEA	*Department of Defense Information Enterprise Architecture Version 1.0*
DoJ	Department of Justice
ESC	executive steering council
GIG	Global Information Grid
IC	Intelligence Community
ISP	Information Support Plan
IT	information technology
LEXS	Logical Entity Exchange Specification
NCDS	*Net-Centric Data Strategy*
NCES	Net-Centric Core Enterprise Services
NCSS	*Net-Centric Services Strategy*

NIEM	National Information Exchange Model
NR-KPP	Net Ready–Key Performance Parameter
NSS	national security system
ODNI	Office of the Director of National Intelligence
RDA	Research, Development and Acquisition
SS	subsurface
UCore	Universal Core
ULEX	Universal Lexical Exchange
VMF	variable message format
XML	Extensible Markup Language

CHAPTER ONE

Objective and Approach

Objective

Universal Core (UCore) 2.0 is a new data exchange framework that uses the Extensible Markup Language (XML) format[1] and that can potentially provide a powerful mechanism for improving interoperability between Department of Defense (DoD) and U.S. government information systems. However, because UCore messages may be large, the implementation of UCore in Department of the Navy networks may cause unintended and negative side effects.

The objectives of this report are to

- describe UCore and its potential benefits and drawbacks
- identify UCore implementation options
- examine implementation issues that may prevent effective use of UCore in Navy networks
- make recommendations to the Navy concerning UCore implementation
- identify UCore implementation guidance in DoD policy and determine whether this policy is coherent and sufficient to guide program managers in executing the DoD data strategy.

Background

UCore development began in April 2007 in response to the need for U.S. government agencies and departments to share information more effectively. DoD policy began citing UCore as early as December 2008. The UCore 2.0 production baseline was released in March 2009, and a DoD Deputy Chief Information Officer (CIO) memo of July 13, 2009, requested information from the services concerning UCore implementation plans.[2]

The Standards, Policy and Guidance Directorate of the Office of the Assistant Secretary of the Navy for Research, Development and Acquisition Chief Systems Engineer (ASN RDA CHSENG) asked RAND to (1) review both the current DoD policy applying to UCore and the technical maturity of UCore and (2) make recommendations concerning its implementation in Navy systems and platforms.

---

[1] For a description of XML 1.0, see Wikipedia, 2010; more information on XML in general, see W3C, 2010.

[2] See U.S. Department of Defense, Deputy Chief Information Officer, 2009.

Approach

The authors reviewed all available technical data on UCore provided by the managers of the UCore Development Team. This consisted primarily of the alpha and beta test reports and materials posted on the UCore website.[3] Additional information about how UCore was being implemented was collected during discussions with the UCore community at the UCore Users' Conference of September 24–25, 2009, and at the National Information Exchange Model (NIEM) National Training Event in October 2009. DoD policy issuances and memos on interoperability and data strategy were reviewed to determine DoD policy on UCore and related requirements for DoD programs. Finally, we conducted a broader examination of subject matter expert publications on policy relating to both information technology (IT) standards and command, control, communications, computers, and intelligence (C4I) interoperability to uncover relevant insights on issues surrounding UCore policy.

Report Outline

Chapter Two of this report describes the UCore data exchange framework, including its core data model, and explores various approaches to UCore implementation. Chapter Three reviews the published results of UCore pilot projects and examines the possible impact of UCore implementation on network bandwidth and cost. Chapter Four assesses the current state of policy on UCore. Chapter Five summarizes the study findings and recommends steps for the Navy (regarding UCore implementation) and for the UCore Development Team (regarding improvements to future UCore pilot projects). These improvements will make the pilot projects more useful for Navy project managers who have to make UCore implementation decisions.

---

[3]   See Universal Core, undated.

CHAPTER TWO

Universal Core Overview

UCore Background

UCore is an XML schema developed and governed by DoD, the Department of Justice (DoJ), the Department of Homeland Security (DHS), and the Office of the Director of National Intelligence (ODNI). It is an information exchange standard designed to facilitate the communication of situational awareness data across the departments by standardizing data syntax and semantics. It consists of a limited but general and flexible data model framework with a semantic context that focuses on *who*, *what*, *when*, and *where*. UCore itself is extensible, which allows users to create more-detailed elements (built on basic UCore elements) that meet their needs. The UCore data exchange framework allows other metadata standards to be "wrapped" in UCore messages as a structured payload. The extensibility and structured payload aspects make UCore different from most other metadata standards.

The UCore effort was initiated in April 2007 with the formation of an executive steering council (ESC) involving DoD and the Intelligence Community (IC). In October 2007, UCore Version 1.0 was released, and DoJ and DHS joined the UCore ESC. There was a beta release of UCore Version 2.0 in September 2008, and the UCore Version 2.0 production baseline was posted in March 2009. UCore supports the data strategy and information sharing goals articulated by the CIOs of four federal departments (DoD, DoJ, DHS, and ODNI).[1] Currently, the UCore development effort is headed by federal co-leads from DoD and DoJ.[2] The UCore current release and documentation reside on the web.[3]

UCore Origins

UCore development did not start from a blank slate. It leveraged a number of previous standards and efforts both internal and external to the four governing departments. DoJ and DHS annually spend $3 billion and $7 billion on IT, respectively. They currently spend $5 million–$7 million annually directly supporting the development of metadata models that improve situational awareness both horizontally across their internal components and vertically

---

[1] See The White House, 2007; U.S. Department of Defense, 2004; U.S. Department of Justice, 2005; Office of the Director of National Intelligence, 2008; and U.S. Department of Homeland Security, undated.

[2] The DoD and DoJ UCore federal co-leads are, respectively Dan Green at the U.S. Navy Space and Electronic Warfare Systems Command and Jeremy Warren at DoJ.

[3] See Universal Core, undated.

between the layers of federal, state, and local agencies. The center of their efforts is NIEM,[4] which is a large, modular, semantic data model that includes both the NIEM core and a dozen domains that containing hundreds or thousands of defined terms. NIEM is a logical model, and, like UCore, it is a flexible data exchange framework.[5] To establish actual standards based on the NIEM model, Information Exchange Package Documents are created that specify the combination of NIEM definitions and schemas, which, respectively, form the semantics and syntax necessary for information exchange between two entities. As one might expect, numerous schemas have been developed to meet the needs of various communities of interest (CoIs) using NIEM, and there has been a great deal of redundancy. To reduce the repeated implementation of NIEM elements and to foster interoperability, DoJ developed the Logical Entity Exchange Specification (LEXS).[6] LEXS is built on the Universal Lexical Exchange (ULEX) schema, which incorporates a structured payload for extensibility and for wrapping other data schemas. Although the LEXS vocabulary is much smaller than the totality of NIEM, UCore is an even more slimmed-down version of LEXS that is also built on the extensible ULEX package structure and is designed to support information exchange between DoD, DoJ, DHS, and the IC.

UCore also leverages a number of other standards:

- DoD Discovery Metadata Specification provides for discovery.
- Intelligence Community Information Security Markings associates security classifications with data using metadata.
- Geography Markup Language instantiates a common understanding of location.
- The Web Ontology Language standard is used as the foundation on which UCore's taxonomy is built.
- UCore's units of measure are consistent with the international trade standards in United Nations Economic Commission for Europe Recommendation 20.

The UCore Data Model and Message Package

The UCore data exchange framework, illustrated in Figure 2.1, provides a message format for defining *things* (using *who*, *what*, *when*, *where*) and the *relationships* between multiple *things* (such as between an *event* and a *person* or *organization*). These tagged data are found in the digest of every UCore package.

The left side of Figure 2.2 describes the contents of the UCore message package, and the right side provides an alternate representation of the message package. The package metadata provide information on the source and timeliness of the data. Of particular note, the metadata include the security classifications of the contents of the package. The UCore security marking function is designed to provide an automated "tear line" capability so that systems operating at a high classification level will be able to remove highly classified data elements from the pack-

---

[4] For more information on NIEM, see NIEM, 2010.

[5] It is important to note that, although NIEM does provide a core vocabulary common to all its Information Exchange Package Documents, individual Information Exchange Package Documents are not necessarily fully interoperable.

[6] For more information on LEXS and ULEX, see LEXS.gov, 2009.

of the other federal departments that participates in NIEM developments might find UCore XML schemas more useful.

If the *base UCore specification* meets the needs of a program, then the program can simply implement UCore. This is most likely to occur when the program needs only to convey basic situational awareness information that fits in the *who, what, where,* and *when* paradigm of the existing UCore data model. By definition, such a program would be interoperable with any other system that uses any form of UCore. However, as previously noted, this is likely to occur only in the case of non-DoD federal department programs that are being designed to implement the schema and semantics of the UCore data model.

The third and fourth options are new approaches to XML data exchange. If these UCore options can be implemented without other drawbacks, they will differentiate UCore from typical XML-enabled data exchange standards.

The third UCore implementation option works in the following way. If a program already has an established XML standard for data exchange, and if the basic UCore elements are not adequate for transmitting the program's data, then, with UCore, it will be possible to wrap a program's XML data by including it in the data to be transported, in the associated XML schema, and in pointers for establishing the legacy schema in an appropriate place in the UCore framework. The data in the legacy XML standard would be embedded in UCore message packets. Any system on the DoD's Global Information Grid (GIG) that could interpret UCore messages would be able to process the original data replicated in the UCore digest but would not be able to interpret the data buried in the legacy XML standard. The unanticipated user would now have access to a minimal amount of data that would at least make him or her aware of the data available in the full message.

The fourth implementation option uses an extended version of the UCore data model. As described earlier, UCore was designed to allow for the development of *UCore extensions*. This feature allows programs to extend UCore's primary data elements by adding characteristics or more-detailed subtypes. UCore extensions could also enable an unanticipated user whose system implemented only the basic UCore data model to understand the parent aspects of the derived data types. This would provide some awareness that a message that used the extended UCore data model contained additional, more-detailed information.

The fifth and sixth UCore implementation options could potentially address the needs of specific CoIs within the Navy if those options do not introduce negative side effects in Navy networks. (This subject is examined later in this chapter.) Due to how recently UCore 2.0 was developed and released, there are no generally acknowledged "established" UCore extensions. However, UCore extensions currently under development are expected to someday become established UCore extensions. In the future, programs will simply adopt one of those UCore extensions. The program's data would be understandable to the CoI for that particular UCore extension and partially understandable to any basic UCore system on the GIG.

Finally, the UCore community of developers has recognized that *UCore extensions are themselves extendable.* In the future, program managers may be able to choose the UCore extension that comes closest to meeting their needs and then extend it to include whatever larger data model is needed by a particular CoI. Conceptually, this would lead to layered data exchange formats in which the higher levels consist of commonly accepted elements that are not allowed to change and in which, at lower levels, programs can innovate with extensions and continue to evolve their data exchange formats. If DoD embraces this model, the development of data exchange formats will accelerate as programs and CoIs experiment with new extensions that presumably would remain backwards compatible with UCore and UCore base extensions. For example, when C2Core is released, it is unlikely to meet all the special command and control data needs of carrier strike groups or subsurface operators (which would likely differ). Rather than either developing their own unique solutions, which could create interoperability challenges, or expanding the effort to develop a common solution, which could impose costs and delays, each platform community could, in theory, extend the C2Core extension of UCore (if it is feasible to implement UCore and C2Core on these platforms and their networks). In this hypothetical example, because both XML data models would be built on the same C2Core foundation, they would be implicitly interoperable at the C2Core level and at the even-higher UCore level.

Traditional Metadata and UCore Interoperability Paradigms

Figure 2.3 is a graphical representation of interoperability based on traditional and metadata-based mechanisms. Systems can communicate directly with one another if they use the same message standards, such as the variable message format (VMF) standard. Message-based interoperability is the oldest and most traditional approach to enabling interoperability between information systems. This method has the drawback that it typically supports point-to-point or single-system-to-single-system information exchanges. In the past, systems developed by different programs or communities typically have developed and used their own message catalogs or standards, so the number of independent message standards has grown faster than the number of information systems. The existence of a large number of independent noninteroperable message standards has made achieving interoperability across a large enterprise, such as DoD, very difficult to accomplish.

A more modern approach is to use XML metadata tags in information transfers between systems, as indicated in Figure 2.3. Systems then communicate using published metadata standards, such as those found in the DoD metadata registry, which is the mechanism consistent with the original DoD data strategy. If two systems can interpret or use the same set of metadata tags, then these systems will be interoperable. However, it is possible that two systems could use different metadata standards to denote the same or similar information. (The possibility that more than one logically and semantically consistent metadata standard may exist and be in use is represented graphically in the figure.) It is also therefore possible that some information systems will be noninteroperable even if the original DoD data strategy was followed.

Indeed, it is possible that DoD information systems that currently use metadata tags may be noninteroperable because they may use distinct metadata standards, as indicated in Figure 2.3. The DoD data strategy directs all programs to register their metadata in the DoD metadata registry, which provides a central metadata standard repository that should enable programs to reuse metadata standards for common data objects. However, if programs do not reuse metadata standards in the registry and instead continually add new metadata to the registry, the probability that the registry will contain redundant or inconsistent metadata will increase. Consequently, the use of metadata standards alone will not guarantee interoperability, and additional measures will be needed. Semantic interoperability problems can still arise even if metadata standards are used.

UCore messages. In addition, the network context of the UCore implementation must be specified to allow a network impact assessment.[5]

Simple UCore messages using only base components of the schema could be small and practical for use over tactical networks. However, given UCore's structured payload, which can carry any data standard, it is possible that live streaming video images might be embedded in UCore messages. On a limited-bandwidth network, those messages might never reach their destination and could bring the network to a halt as it repeatedly tries to deliver messages or message packets. Alternatively, UCore messages could be implemented such that large data files only appear as attachment links in UCore messages, thus allowing the system to receive and transmit UCore messages while selectively choosing to pass (or not pass) the bulk of the data. Even when UCore systems are fully realized and concrete measurements of message size are produced, there will be a large range of variability depending on the implementation.

XML Compression

A solution that may enable the use of UCore on limited-bandwidth networks is XML compression. XML is mostly text, which is very suitable for compression, and several XML compression schemes have been developed by commercial firms.[6] The Network Enabled Command Capability program at the Defense Information Systems Agency has evaluated several possible XML compression schemes for use in DoD networks (Net-Enabled Command Capability, 2008). One particular XML compression scheme, Efficient XML, can compress XML messages by a factor of 30 to 600, depending on the size of the original XML message. Large messages are compressed by a greater factor than small messages. RAND has examined the use of XML compression in tactical networks and has found that compression can reduce the impact of XML messaging on tactical networks significantly and may make it feasible to use XML messaging for some message types, such as blue force tracking messages (Gonzales et al., 2010a). The MITRE Corporation is also beginning to explore the use of compression with UCore specifically (MITRE Corporation, 2009). XML compression of UCore messages will probably reduce the network bandwidth required to support UCore messaging, but more data on UCore message sizes are needed to determine whether the bandwidth reduction factor provided by compression is sufficiently large to make UCore messaging feasible over tactical networks.

UCore Cost Concerns

A separate but not insignificant concern is the cost implications for Navy acquisition programs of implementing UCore. Credible cost information is needed in order for a program manager to make an informed decision about whether to implement UCore.

UCore implementation costs, like UCore bandwidth issues, are difficult to predict. First, there are very few hard data on costs. UCore efforts have primarily been small pilot efforts for technology demonstration within larger programs with limited cost visibility. These pilot

---

[5] E.g., is the supporting network fiber-optic, satellite communications–based, or tactical data link–based?

[6] For example, see XML Liquid Technologies, undated.

efforts have focused on demonstration with minimal associated time and labor costs, and they have not focused on meeting rigorous system requirements (which would entail greater costs).

Second, the implementation scheme chosen for a particular UCore application may greatly affect implementation cost or the cost of transitioning from current messaging schemes and standards to UCore. Wrapping an established data standard in UCore message packages, translating or mimicking similar data standards (such as LEXS), or choosing a limited implementation of essential UCore elements (as found in most of the pilot efforts) is likely to entail very limited costs. But costs associated with upgrading Navy data models to UCore and with populating associated UCore metadata fields may be significant, especially for legacy systems for which formal metadata do not exist. Some cost estimates indicate that implementing UCore with C2Core may add 10 percent–25 percent to the cost of current practices.

Findings

The cost of implementing UCore will be highly dependent on the implementation approach used. Wrapping established data standards in UCore messages may be generally affordable, but creating UCore extensions may be very expensive. Furthermore,

- UCore pilot projects are not well documented and do not contain sufficient technical information to allow acquisition program managers to make informed UCore implementation decisions.
- Detailed technical data on UCore implementation options, including information on UCore message sizes, are not yet available.
- Bandwidth requirements for the UCore implementation options are not known, even for a minimal implementation of UCore that includes only the core UCore data model.
- Analytical results from RAND studies that examined XML messaging indicate that implementation of UCore on information systems that use Navy tactical or satellite networks may not be feasible without the use of XML compression.

CHAPTER FOUR

Policy on UCore

Most DoD policy on data strategy was developed before UCore existed and therefore does not mention it. A number of more-recent DoD CIO policy documents mention UCore but do not specify how UCore should be used. Several new DoD policies currently under development and review should provide more-specific guidance on UCore.

Data Strategy Policy

The primary sources of policy on DoD data strategy are the following:

- DoD Directive (DoDD) 8320.02
- DoD 8320.02-G (guidance issued by the DoD CIO under DoDD 8320.02 authority)
- *The Department of Defense Net-Centric Data Strategy* (NCDS)
- *The Department of Defense Net-Centric Services Strategy* (NCSS) (issued by the DoD CIO and cited by Chairman of the Joint Chiefs of Staff Instruction [CJCSI] 6212.01E).

As all of these sources were issued before UCore was developed, so none mentions UCore.

Recent Interoperability Policy

Currently, CJCSI 6212.01E (released December 15, 2008) is the only DoD-wide adjudicated policy that mentions UCore. This policy establishes a new Net Ready–Key Performance Parameter (NR-KPP) to assess the interoperability and supportability of DoD IT and national security systems (NSS). The NR-KPP consists of five elements: solution architectures, data and service strategy, technical standards, information assurance, and supportability. The data and services strategy element requires that data and services be visible, accessible, understandable, secure, and interoperable. Under the instruction's "Data and Services Must Be Interoperable" section, UCore is mentioned in the following sentence: "Semantic, structural and security artifacts for data sharing shall be derived from the Universal Core (UCore), domain cores (e.g., C2Core), COIs, and other data standards in accordance with reference v" (Joint Chiefs of Staff, 2008, p. E-8), where reference v is the NCDS. This statement is ambiguous. It can be construed as policy requiring the use of UCore whenever applicable, but NCDS issuance precedes UCore and does not mention it. It can also be more broadly interpreted as suggesting that UCore is just one of many data standards that can be used to meet interoperability requirements in NCDS. CJCSI 6212.01E provides no further guidance for interpreting that statement. Elsewhere in CJCSI 6212.01E, UCore is twice mentioned purely as an example.[1]

Additional Memoranda and Informal Policy

The DoD CIO has issued memoranda and policy that address UCore. *Department of Defense Information Enterprise Architecture Version 1.0* (DoDIEA) mentions UCore in passing twice:

> DoD programs providing IT capabilities must also adhere to applicable DoD CIO established global standards such as the Universal Core information exchange schema and use, where appropriate, Core Enterprise Services provided through the Net-Centric Core Enterprise Services (NCES) program. Additionally, DoD IT leverages the shared common computing and communications infrastructure of the Global Information Grid (GIG). Non-GIG IT includes stand-alone, self-contained, or embedded IT that is not and will not be connected to the enterprise network. (U.S. Department of Defense Chief Information Officer, 2008, p. 1)

> Semantic vocabularies shall re-use elements of the DoD Intelligence Community (IC)-Universal Core information exchange schema. (U.S. Department of Defense Chief Information Officer, 2008, p. 11)

The first case cites UCore as an example, and the second refers only to the semantics of UCore and presumably is not applicable to the UCore syntax. More recently, the DoD CIO and DoD Deputy CIO have issued memos specifically on UCore. *Department of Defense (DoD) and Intelligence Community (IC) Initial Release of Universal Core (UCore)*, the joint DoD CIO and Associate Director of National Intelligence, CIO, memo of April 17, 2008, describes UCore's early form and states only that UCore is consistent with the DoD data strategy in DoDD 8320.02 and 8320.02-G. On July 13, 2009, the DoD Deputy CIO issued a memo requiring that DoD departments respond to a survey on UCore implementation plans that made these statements:

---

[1] From Joint Chiefs of Staff (2008, pp. E-8–E-9):

> Verification of requirements documentation compliance with the DOD Net-Centric Data Strategy and DOD Net-Centric Services Strategy (references v and w) will be accomplished through the analysis of the sponsor-provided architecture and verification products with accompanying text detailing the program's compliance strategy. Documentation (in solution architecture products or other forms) must clearly identify all net-centric services and data, including any adopted from the Ucore (reference ww), Domain Cores and COIs.

From Joint Chiefs of Staff (2008, p. E-9):

> CPDs [Capability Production Documents] and Milestone-C ISPs [Information Support Plans] (including ISP annexes for incrementally fielded capabilities) will include the Logical Data Model (OV-7) and the Physical Schema (SV-11) if the system being described shares any internal data with external systems. If the system accesses shared data from an external system, then the document may point to the external system's OV-7 and SV-11 (if available) by reference. The SV-11 should include any metadata namespace (examples include XML or HTML [Hyper Text Markup Language] schemas) in the DOD Metadata Registry that documents data standards used by the proposed system, data derived from data models or other standards, such as the UCore.

DoD Components must begin planning efforts for its [UCore's] adoption by programs of record and in the transition of legacy environments. (U.S. Department of Defense Deputy Chief Information Officer, 2009, p. 2)

Guidance will be updated to include the use of UCore, domain common cores, and COI share vocabularies. Although austere environments present some unique challenges, UCore should be used to the maximum extent practical as work continues to enhance performance in these environments. (U.S. Department of Defense Deputy Chief Information Officer, 2009, p. 1)

The memo obviously encourages the use of UCore across all DoD IT and NSS, but it does not mandate its use.

Further guidance mentioned in the memo does not yet exist. Clay Robinson of the DoD CIO's office noted at the September 2009 UCore Users' Conference that DoD 8320.02-G is being rewritten to address UCore. It is also our understanding that DoDD 4630.05, *Interoperability and Supportability of Information Technology (IT) and National Security Systems (NSS)*, and CJCSI 6212.01E may be under review or in the process of being rewritten to make their separate versions of the NR-KPP consistent. At minimum, CJCSI 6212 is under construction to provide additional guidance on the NR-KPP; it may very well also include additional UCore guidance.

UCore in Policy: Standards Definition Versus Data Model

One of the challenges presented by UCore is that some organizations treat it as standards definition while others consider it a logical data model. In reality, it is neither. In the acquisition community, especially at the systems or computer engineering levels, UCore is nothing more than a standard or specification for encoding data. As with all standards chosen by a program, the pros and cons—such as capability, general acceptance, technical risk, and cost—must be compared with those of other standards that might meet program needs. Although the selection of a standard will impose some basic constraints on the system, given the standard's particular limitations, simply selecting a standard does not define how and to what extent that standard is used by a program. This is particularly true with UCore because of the large number of available implementation approaches discussed in Chapter Two.

In contrast, developers of DoD data architectures, managers of DoD C4I systems, and generators of related DoD policy are likely to view UCore as a logical model. Data models require both standards for establishing the syntax that facilitates the transfer of the data and a framework for common semantic definitions such that everyone understands what the communicated information means. A fully conceived logical data model may also include a governance system for clarifying or extending the model's semantics. The DoJ/DHS NIEM mentioned in Chapter Two is an example of an extensible data exchange framework that is not a logical data model but rather a framework that can support multiple data models. Although the construction and extension of NIEM semantics are controlled through high-level design guidelines, the syntactic implementation of NIEM occurs through many different standards that are established through the model's information exchange package documentation.

UCore is neither just a standard nor a single logical data model; it shares elements of both but it is also not a complete data model. UCore utilizes XML standards. However, XML standards are very general and support multiple data models. In addition, that UCore leverages numerous commonly accepted semantic standards, as detailed in Chapter Two. Recognizing that one XML standard or data model could never meet everyone's needs, UCore designers made it extensible in a number of ways. However, only the technical rules for extension exist: Logical extension guidelines and governance do not currently exist in the DoD or across the federal departments that govern the UCore standard.

Policy approaches for establishing IT standards are different from those for adopting a new data exchange or a new framework data model. One concern about DoD policy on UCore is that policy writers and program managers will interpret UCore as a standard or as a single complete data model, when in reality it is neither. This could lead to a great deal of confusion.

Findings

This chapter demonstrates the following:

- Current policy does not mandate or require the use of UCore to meet DoD data strategy and broader interoperability and supportability policy.[2]
- There is no specific UCore implementation guidance.
- Recent issuances from the Chairman of the Joint Chiefs of Staff and from the DoD CIO suggest that the DoD is moving toward mandating UCore.
- UCore policy can be written and interpreted as if UCore is an IT standard or a single complete data model, but, in reality, it is neither.

---

[2] As previously mentioned, current policy is not clear in this regard. The authors have confirmed by consulting with officials in the Assistant Secretary of Defense for Networks and Information Integration that current DoD policy does not mandate UCore.

CHAPTER FIVE

Conclusions and Recommendations

UCore Has Promise

On the surface, UCore is simply another XML-based data standard available to the DoD acquisition community for establishing interoperability between DoD information systems. However, the data-standard wrapping and extensibility capabilities inherent in UCore give it the potential to significantly improve interoperability between DoD information systems and provide a capability that is needed to realize the promise of the DoD data strategy.

An important aspect of the DoD data strategy is understandability, which requires both semantic and syntactic elements to achieve a working data exchange framework. Widespread adoption of a core set of common syntactic or XML standards, such as UCore, will improve understandability, especially for the unanticipated user. However, UCore is not a complete data model and by itself, as it exists today in UCore version 2.0, will not meet the interoperability needs of the Navy or the broader DoD. Extensions to UCore, such as C2Core, are needed to make the schema more useful for Navy and other DoD users. The extensibility of UCore makes this increased usefulness possible but also presents opportunities for compatibility and integration problems because UCore is complex and provides several extension mechanisms.

Recognizing the potential benefits of UCore for DoD interoperability, the DoD CIO is advocating for accelerated adoption of UCore by acquisition programs through policy directives, instructions, and memorandums, as described in Chapter Four.

UCore Policy Is Immature

Most existing DoD policy relevant to the DoD data strategy does not mention UCore, and new DoD policy (especially CJCSI 6212.01E) provides ambiguous and confusing guidance on UCore implementation requirements. Establishing a clear-cut DoD policy on UCore is complicated by the fact that it is neither a simple XML standard nor a single complete data model that can meet all of DoD's interoperability requirements. Since UCore fits neither paradigm completely, stakeholders in UCore policy are likely to interpret UCore differently—as a standard or a data model or, more correctly, as a flexible data exchange framework that can support multiple data models. This creates policy challenges, since the authority to generate policy and guidance on UCore depends on how UCore is defined and understood.[1]

---

[1] For an in-depth exploration of IT roles and responsibilities in the DoD, see Gonzales et al., 2010b.

DoD should promote the adoption of UCore by addressing the technical issues and cost concerns raised in this report, increasing the visibility of UCore, expanding the community of UCore users, and refining the guidance related to UCore extensions and implementation. UCore should be improved, and more-robust UCore implementations should be tested before its use is mandated in DoD policy.

UCore Pilot Projects

A limited number of UCore pilot projects have been conducted to date. These UCore pilot projects are not well documented, and there is not sufficient technical information to help acquisition program managers make informed UCore implementation decisions. Furthermore, detailed technical data on UCore implementation options, including information on UCore message sizes, are not yet available, even for only the minimal implementation of UCore. Therefore this body of evidence cannot be used to assess the impact of UCore on Navy networks or to assert that UCore can be implemented effectively and without risk on Navy networks with bandwidth limitations.

UCore Implementation Options and Risks

There are several ways to implement UCore. Some of these implementation options are feasible only if extensions to UCore are available. Other options can be implemented more quickly if existing XML schema are wrapped in UCore. Limited guidance and supporting technical information that describe the potential advantages and disadvantages of these implementation options, their costs, and their impact on the Navy networks are available.

The lack of UCore implementation guidance and the limited understanding of technical and cost implications of its adoption are a source of concern in the DoD acquisition community, especially in the light of UCore's flexibility and the variety of implementation methods available, which greatly complicate the technical and cost assessment of UCore systems.

XML messaging is typically verbose, and UCore is no exception. Additionally, the extensibility and wrapping inherent in UCore messages make the assessment of network impact even more difficult because there is significant potential variability in UCore message sizes. Finally, the cost of implementing UCore will vary greatly depending on specific implementation choices. These technical issues need to be understood and addressed before UCore implementation can be mandated.

Recommendations

In light of these findings, we make the following recommendations:

- The Navy should not require widespread UCore implementation at this time.
- DoD should undertake additional UCore piloting efforts to quantify potential system and program impacts. It should

  – Capture UCore vocabulary size, message sizes, system processing speeds, and network bandwidth consumption.
  – Model and document program UCore implementation costs for specific UCore implementation options.
- Any future UCore requirements or mandates should not originate ab initio from policy but should rather be informed first by the evaluation of technical and cost data from well-documented future pilot projects.
- Guidance for setting UCore requirements in program Initial Capabilities Documents should include an analysis of alternatives that weighs interoperability benefits against bandwidth impact. For example, systems that will regularly interact with DoJ DHS/IC systems clearly need to implement UCore or an equivalent, but tactical-edge systems with limited bandwidth might be forbidden to use UCore, and headquarters intelligence databases attached to wideband networks that accumulate data from the tactical edge might be allowed to generate UCore messages for passing information up the chain of command on high-bandwidth networks.
- The Navy should permit UCore implementation if (1) the program depends only on high-bandwidth networks or (2) UCore will not degrade or otherwise affect required real-time performance requirements for the system or system of systems to which the subject system belongs.
- The Navy should develop guidance for preferred UCore implementation approaches that accounts for different Navy networks and operational environments.
- DoD should develop a new, streamlined version of Ucore (UCore 3.0) that contains a small and extensible core vocabulary for the UCore core primitive data frameworks for "who" and "what."

Future Work

Related RAND research for the Director of Joint Interoperability, Under Secretary of Defense for Acquisition, Technology and Logistics, has shown that XML compression can greatly reduce the size of XML-tagged messages. This may enable UCore messaging to be used on networks that have bandwidth limitations. Further research is required both to determine how effective XML compression is when applied to UCore messaging and to determine which Navy network types could support compressed UCore messaging.

One of the greatest long-term potential benefits of UCore adoption will be the ability of different CoIs to independently develop improved data tagging using UCore or a UCore extension while remaining interoperable (to a certain degree). However, to realize that capability, the UCore data model needs to avoid becoming a centrally managed affair—a situation that could inhibit innovation (as has occurred in the case of the NIEM). DoD and the Navy should develop alternative ways of managing the UCore data model that encourage program experimentation while ensuring that bandwidth issues and implementation costs are addressed.

Bibliography

ASD (NII)—*see* U.S. Department of Defense, Assistant Secretary of Defense (Networks & Information Integration).

Briscoe, Bob, Andrew Odlyzko, and Benjamin Tilly, "Metcalfe's Law Is Wrong," *IEEE Spectrum*, July 2006, pp. 26–31.

Committee to Review DOD C4I Plans and Programs, Commission on Physical Sciences, Mathematics, and Applications, National Research Council, *Realizing the Potential of C4I: Fundamental Challenges*, Washington, D.C.: National Academy Press, 1999.

Gonzales, Daniel, Sarah Harting, Rima Bitar, Angel Martinez, and Brien Alkire, *Supporting the Warfighter at the Edge: Can Tactical Edge Networks Support Service Oriented Architectures?* RAND Corporation, 2010a. Not available to the general public.

Gonzales, Daniel, Carolyn Wong, Eric Landree, and Leland Joe, *Are Law and Policy Clear and Consistent? Roles and Responsibilities of the Defense Acquisition Executive and the Chief Information Officer*, Santa Monica, Calif.: RAND Corporation, MG-958-NAVY, 2010b. As of September 20, 2010:
http://www.rand.org/pubs/monographs/MG958.html

Joint Chiefs of Staff, *Interoperability and Supportability of Information Technology and National Security Systems*, CJCSI 6212.01F, December 15, 2008.

LEXS.gov, homepage, 2009. As of September 20, 2010:
http://www.lexs.gov/

Libicki, Martin, James Schneider, David R. Frelinger, and Anna Slomovic, *Scaffolding the New Web: Standards and Standards Policy for the Digital Economy*, Santa Monica, Calif.: RAND Corporation, MR-1215-OSTP, 2000. As of September 20, 2010:
http://www.rand.org/pubs/monograph_reports/MR1215.html

MITRE Corporation, untitled briefing, UCore Users' Conference, McLean, Va., September 24 and 25, 2009.

Net-Enabled Command Capability, *XML Compression Technology Maturity and Readiness Analysis, Assessment Report Version 1.0*, Defense Information Systems Agency, October 16, 2008.

NIEM, homepage, last updated September 3, 2010. As of September 20, 2010:
http://www.niem.gov/

Office of the Director of National Intelligence, *United States Intelligence Community Information Sharing Strategy*, February 22, 2008.

UCore v2.0 Development Working Group, *Universal Core (UCore) v2.0 Alpha Consolidated Pilot and Evaluation Report*, August 26, 2008.

———, *Universal Core (UCore) v2.0 Beta Phase Final Report*, March 31, 2009.

Universal Core, homepage, undated. As of September 20, 2010:
https://ucore.gov/ucore

U.S. Department of Defense, *Interoperability and Supportability of Information Technology (IT) and National Security Systems (NSS)*, DoDD 4630.05, May 5, 2004.

———, *Data Sharing in a Net-Centric Department of Defense*, DoDD 8320.02, December 2, 2004, certified current as of April 23, 2007.

U.S. Department of Defense, Assistant Secretary of Defense (Networks & Information Integration), *DoD Command & Control Implementation Plan V1.0*, October 1, 2009.

U.S. Department of Defense, Assistant Secretary of Defense (Networks & Information Integration), and Department of Defense Chief Information Officer, *Guidance for Implementing Net-Centric Data Sharing*, DoD 8320.02G, April 12, 2006.

U.S. Department of Defense Chief Information Officer, *DoD Net-Centric Data Strategy*, May 9, 2003.

———, *Department of Defense Net-Centric Services Strategy: Strategy for a Net-Centric, Service Oriented DoD Enterprise*, May 4, 2007.

———, *DoD Information Enterprise Architecture Version 1.0*, April 11, 2008.

U.S. Department of Defense Chief Information Officer and Associate Director of National Intelligence, Chief Information Officer, *Department of Defense (DoD) and Intelligence Community (IC) Initial Release of Universal Core (UCore)*, April 17, 2008.

U.S. Department of Defense Deputy Chief Information Officer, *Universal Core (UCore) Guidance in Support of Enhanced Information Sharing Memorandum*, July 13, 2009.

U.S. Department of Homeland Security, *One Team, One Mission, Securing Our Homeland: U.S. Department of Homeland Security Strategic Plan, Fiscal Years 2008–2013*, undated.

U.S. Department of Justice, *LEI SP: United States Department of Justice Law Enforcement Information Sharing Program*, October 2005.

W3C, "Extensible Markup Language (XML)," web page, last modified September 6, 2010. As of September 20, 2010:
http://www.w3.org/XML/

The White House, *National Strategy for Information Sharing: Successes and Challenges in Improving Terrorism-Related Information Sharing*, Washington, D.C., October 2007.

Wikipedia, "Semantic Web," web page, last modified September 19, 2010. As of September 20, 2010:
http://en.wikipedia.org/wiki/Semantic_web#Purpose

XML Liquid Technologies, "XML Compression," web page, undated. As of September 20, 2010:
http://www.liquid-technologies.com/Xml-Compression.aspx